Something to Talk About

Louise Mayes Durall

Bloomington, IN  Milton Keynes, UK
 authorHOUSE®

AuthorHouse™
1663 Liberty Drive, Suite 200
Bloomington, IN 47403
www.authorhouse.com
Phone: 1-800-839-8640

AuthorHouse™ UK Ltd.
500 Avebury Boulevard
Central Milton Keynes, MK9 2BE
www.authorhouse.co.uk
Phone: 08001974150

First published by AuthorHouse   5/8/2007

ISBN: 978-1-4259-8355-0 (sc)

Library of Congress Control Number: 2006910905

Printed in the United States of America
Bloomington, Indiana

This book is printed on acid-free paper.

Dedication

After I retired I returned to Greenport, N.Y. to assist my mother Pauline Strange until she passed. She constantly reminded me to reach for the sky. A certificate of Certification from the State University of Farmingdale, N.Y. qualified her to become a Family Day Care Mother in the Village of Greenport, N.Y. She taught children who needed help in school, at home, or in church. They later sent her letters of thanks when they graduated from high school, entered college or finished college. They thanked her for teaching them that there is "always" room and "never" too late to improve your education.

I volunteered with the Outreach in Southold, N.Y. to stay busy and assist young people interested in learning about our weather. It was having something to talk about. An interested high school student graduated from high school, and upon completion of college became an employee of the National Weather Service. I am hoping this book will benefit in assisting other young people interested in our weather.

I dedicate this book to all above and my deceased son Austin Durall, who was highly thought of as an Electronic Computer Repairman in the United States Air Force.

Acknowledgements

I am very grateful to my daughter Sheryl a 4th grade school teacher and my granddaughter Georgette, a Document Production Manger in assisting me with the computer knowledge used in the completion of this book. I must also include my Granddaughters Monique and Angela for their encouragement. My granddaughter Angela and her son Dante have inspired me to write a Children's Book about the weather.

I am also grateful to my co-workers at the National Weather Service and my other friends for their interest in my career, and proving support over the many years. I apologize for not listing any names individually.

OBSERVING AND FORECASTING THE WEATHER

The Technology of Observing and Forecasting Weather became a demanding discovery by the National Weather Service. Weather is described as the condition of the atmosphere. Conditions of the atmosphere are barometric pressure, temperature, humidity, winds, precipitation, and clouds. The information can be shown on a weather map which includes the direction and velocity of the winds over an area, for a given time. If a sudden change in the wind direction should occur after the forecast is given; the forecaster will immediately issue a new forecast. The reason for the sudden change in the wind direction is influenced by Mother Nature; and she cannot be disputed.

Table of Contents

My family blood lines are rooted in Guyana and Venezuela. That explains the color of my eyes. Which I'm so frequently ask about.

When I speak of my career I've been asked to write a book. Writing is having something to say, about what you know, if not, it would be forgotten. During my years of retirement, I've learned that a passion for reading enhances the mind. On that note, I'll venture into writing this book.

On June 1, 1986, I received a Certificate of Service from the Department of Commerce. The certificate was presented to me by the Meteorologist in Charge of the National Weather Service Forecast Office at the Rockefeller Center office. It was awarded in recognition of 35 years of dedicated service rendered to the Federal Government   of the United States. The certificate has a gold seal, and is signed by the Secretary of Commerce and the Administrator of the National Oceanic and Atmospheric Administration.

I am the first person of color who has been employed in the New York City area, by the Department of Commerce with all of these position titles: Weather Data Editor, Meteorological Technician/Forecaster Aid, Upper Air Weather Observation Specialist and Climatologist for the National Weather Service at the Rockefeller Center office.

My 35 years of employment with the Department of Commerce was in the following government agencies: Civil Aeronautics Agency, Federal Aviation Administration, National Oceanic and Atmospheric Administration and the National Weather Service.

My career consisted of working predominately with male employees. I consider myself very fortunate to have worked among respectful men, who showed me respect and were careful not to use inappropriate language in my presence.

My potential was to open doors of success for me and other people of color to follow. I had a strong determination to set an example of what a person can do, if they put their mind to it. I indulged hook, line and sinker, not because I had to, but because I wanted to. Perseverance, courage, capability, integrity, and self esteem also contributed to my accomplishing all that I could. (A picture is worth a thousand words).

The Western Union Telegraph Company was my first place of employment after graduation from

Greenport High School, in Greenport, Long Island, N.Y. I was one of the first people of color employed as a Teletype and Telegraphic Operator at their main office in New York City. I passed the Pennsylvania State University Entrance Examination which was a requirement for employment, and I knew how to type.

At the Western Union School of Teletype and Telegraphic Technology; I learned to type on the teletype machine and also learned the knowledge of coding and sending messages to be transmitted for distribution by teletype. The messages sent by teletype to be transmitted for distribution consisted of printed tapes. The training included routing messages received over telegraph to be transmitted for distribution by teletype. I learned to code, send and receive messages of perforated tapes received from the Mux machine to be transmitted for distribution. I also learned to receive, read and send messages in Morse code.

During my employment with Western Union I got married and moved to California where my husband was stationed in the military service. I took a leave of absence to continue my employment with Western Union and remained in California until my husband was ready to be shipped overseas, and I returned to New York.

When my husband returned home after honorably serving the United States Navy-Coast Guard; he was entitled to wear the following ribbons: American Area, Asiatic-Pacific Area, Philliphine Liberation Campaign Ribbon, World War II Victory Ribbon, American Defense and Amphibious Force Insignia and 2 Bronze Stars. We were blessed with three children, and I returned to work.

I became a Federal Government employee with the Supervisor of Shipbuilding under the Civil Aeronautics Agency in the Department of Commerce. My position title was Semi- Automatic Teletype Operator. I typed messages on the teletype and transmitted the digital readings over the teletypewriter for distribution. I also sent messages received over telegraph to be transmitted for distribution by teletype to military destinations, under the supervision of the United States Navy Department. When a vacancy became available for Clerk Typist with Bookkeeping within the same agency I applied for the position, and was accepted at an increase in salary.

MY CAREER BEGAN WITH THE STUDY METEOROLOGY

AS: WEATHER DATA EDITOR

My experience in Weather Technology began with the U. S. Weather Bureau under the Federal Aviation Administration in Hangar 11 at JFK International Airport. The title of my position was Weather Data Editor and required working rotating shifts. My position description consisted of; my receiving and editing the hourly weather reports received from telegraph to be transmitted by teletype. It included coding the perforated tapes for distribution by teletype. It also included editing the hourly weather reports received from telegraph to be transmitted by teletype for distribution to the airports.

"Synoptic Meteorology" was included in the position description. I collected many simultaneous observations that covered large areas from telegraph and transmitted the observations for distribution to central points by teletype. The hourly weather observations were entered on weather maps and interpreted through the Meteorologists' knowledge of all the physical data that were available.

When the weather maps were plotted the symbols and figures were drawn in a way that the changes could be predicted by countries throughout the world. The weather reports were sent in international code. Once the data was plotted the charts were analyzed by forecasting experts, who drew isobars to show lines of equal pressure. Then they marked position of fronts and showed precipitation areas. The forecasters prepared the forecasts and weather bulletins from these analyze charts.

The pilots of the airplanes at JFK and LaGuardia Airports would come into the Control Towers Briefing Room at Hangar 11 to be briefed on the latest weather reports, before they "took-off" on their flight. The pilots would receive scheduled weather reports and forecasts every half hour while in flight. They insisted on the latest reports, because planes move very quickly and the weather changes continuously. Weather does not stand still. Sometimes the wind direction would change suddenly due to the influence of Mother Nature and the forecasters would immediately prepare a new forecast. The forecasters were constantly on the alert and prepared to meet the pilots' requests for the latest forecasts. The passengers aboard the planes were referred to as "souls" aboard the planes.

One day at JFK International Airport, a weather plane was parked outside Hangar 11. It had a scary dismal gray color, built large and very rugged looking. I decided to take a tour of the plane to investigate the inside. From the inside of the plane and while looking out the window I sensed a feeling of danger. The weather planes are built to survive most of the damage that might occur during the flight into areas of violent and severe storms. The pilots would fly the weather planes into the eye of the hurricanes and took observations pertaining to the severity of the storm. The observations were   included in the weather reports.

A time came when the U.S. Weather Bureau, under the Federal Aviation Administration discontinued their occupancy at JFK International Airport, and a Reduction in Force took place. My name was included in the Reduction in Force, because I didn't have enough government seniority to compete. Therefore, I had to seek other employment.

At that time a Stock Market on Wall Street was interviewing teletype operators for employment. I applied for the position and was accepted. In the meantime I acquired Veterans Preference as an employee of the Federal Government. This happened after my husbands' death from cancer. As a result, I continued my employment with the Federal Government in the Department of Commerce.

Two male co-workers from JFK International Airport Hangar 11 were interested in working with the Stock Markets on Wall Street as Teletype Operators. I assisted them in reference to the questions that were asked on my interview. The information helped to prepare them for their interview, and they were hired.

CONTINUATION OF MY CAREER

AS: WEATHER DATA EDITOR

The Weather Bureau became known as the National Weather Service, a branch of the National Oceanic and Atmospheric Administration, in the Department of Commerce located at the New York University Technology Building 2, in the Bronx.

My position title continued as Weather Data Editor with rotating shifts. I was responsible for receiving and editing hourly weather reports that were received over telegraph to be transmitted for distribution on the teletype machine. I also analyzed the present weather conditions, and extrapolating these conditions and trends into the future, to be transmitted for distribution by teletype such as; information received from the 24 hour forecasts, the 24 hour aviation forecasts, the 5 day forecasts, weather conditions at the airports, special bulletins for weather maps, reports of temperature and precipitation, warnings on storms, farms, industry, frost, severe weather reports. I completed the temperature map that appeared in the New York Times newspaper.

Louise Mayes Durall

When the National Weather Service Forecast Office moved from the New York University Technical Building 2, in the Bronx to the Rockefeller Center office, I received the following letter signed by the Meteorologist-in-Charge:

"Dear Mrs. Durall:

I would like to express my thanks and commend you for help in solving the communications problem between Pittsburgh, PA and this office. The New York Weather Service Forecast Office had not received the public forecast for several months, due to the incorrect communications procedures and faulty equipment.

Your tenacity and working with the communicators at Pittsburgh eliminated the possibility of faulty coding of the teletype messages. The technicians of the telephone company were then able to trace any additional trouble in the switching mechanism in the Overlay Circuit at Pittsburg. A copy of this letter is being placed in your personnel file".

The success in helping to solve the communications problem was due to my doing more than I needed to, not because I was asked, but because I want to. My ability to succeed successfully as a volunteer was due to the excellent training I received at the Western Union School of Teletype and Telegraphic Technology.

The invention of the telegraph and teletype machines ended the theory of Mythology, also the paid and unpaid volunteer's cooperation in supplying the government with weather information. The new technology made it possible to receive and send all types of messages on a current basis. My knowledge of the telegraph and teletype machine became the tool necessary to achieve the goal in my career.

I arrived at another milestone in my career, when the National Weather Service assigned me to the National Weather Service Technical Training Center in Kansas City, Missouri. Tornados occur quite often in Kansas City, Missouri. None occurred during my training. I completed the training and received a certificate as Basic Meteorological Technician. I also received a certificate for Writing and Broadcast Seminar both certificates were signed by the Director of the National Weather Service.

I returned to the National Weather Service Forecast Office at the Rockefeller Center office and was presented a "Certificate of Authority" by the National Weather Service to take weather observations. The "Certificate of Authority" stated: "This is to certify that Louise A. Durall is qualified and authorized to take weather observations subject to the conditions stated in the National Weather Service Operations Manual." I also received a certificate to become an associate member of the Association of Air Traffic Specialists, serving the flying public in the field of flight service signed by the Executive Director.

CONTINUATION OF MY CAREER

AS: METEOROLOGICAL TECHNICAN/ FORECASTER AID

The journey into my career continued to move upward due to my outstanding training at the National Weather Service Technical Training Center in Kansas City Missouri. I learned the technology needed to successfully succeed in the Technology of Meteorology. The evaluation of my work performance was rated as exceeded and my expertise considered valuable to the forecaster and public. My position title became Meteorological Technician and Forecaster Aid.

The National Weather Service assigned me to JFK International Airport Control Tower. My position title was Meteorological Technician and Forecaster Aid with rotating shifts. I maintained the records of all incoming and outgoing planes, including runway and flight conditions that could be valuable to the controller. Under the supervision of the Tower Controller I transmitted the latest weather predictions and forecasts to the pilot at all major airports by teletype.

As an air safety precaution, major airports have stations operated by   the Federal Aviation Administration or the National Weather Service, to give pilot's complete flight information. Accuracy was considered   extremely important to the pilot for safe flying. It was also important to the pilot to be informed of how to avoid dangerous flying conditions. The pilots insisted on the latest reports, because the planes move swiftly, and the weather is constantly changing.

Air Safety Precautions and Flight Information given to the pilots at JFK International Airport Tower were:

1. Thunderstorms are sure signs of violently unstable air..........Give these storms a wide berth.

2. Shallots and clouds towering upward is a great force of energy, and indicate strong updrafts and rough turbulent air..........Stay clear of these clouds.

3. Fair weather cumulus clouds often indicate bumpy turbulence beneath the cloud and in the clouds. The cloud tops indicate the approximate upper limit of convection.......... Flight above is usually smooth.

4. Stratus formed clouds indicate stable air..........Flight generally will be smooth, but low ceiling and visibility might require

flying on Instruments...Instrument Flight Regulations.

5. Even in clear weather, you have some clues to stability such as; if temperatures remain unchanged or decreases only slightly with altitude..........The air tends to be stable.

When my assignment was completed at JFK Airport Control Tower, the National Weather Service assigned me to the LaGuardia Airport Control Tower. My position title continued as Meteorological Technician and Forecaster Aid with working rotating shifts. Under the supervision of the Tower Controller, I would radio broadcast the latest forecasts with the latest weather conditions that were made available to the pilots.

Radio broadcasts were made from LaGuardia Control Tower, to the pilots of large planes, crossing the Atlantic Ocean. The private and commercial aircraft also received radio broadcasts. All airports tuned in on the radio broadcasts from LaGuardia Control Tower to get the latest reports and weather conditions when the National Weather Service Forecast Office phones were busy. More people were traveling by air every year, therefore great demands were expected from the weather people.

Air Safety Precautions and Flight Information given to the pilots at LaGuardia Airport Control Tower were:

1. The upper air weather reports are very important to the pilots because his speed

over the ground depends upon the winds in the upper levels of the atmosphere.

2. At night the visibility is determined by the distance light can be seen, by the pilot.

3. The visibility is very important to the pilot it is the distance that an object is barely seen with the eyes.

4. If the pilot knows the visibility at an airport where he is going to land, he will know how far ahead the runway can be seen.

5. In cloudy weather, when a pilot is ready to land he must know how high he should be above the ground, in order to see the runway.

At the completion of my assignments at JFK and LaGuardia Airport Towers the National Weather Service again assigned me to the National Weather Service Technical Training Center, in Kansas City Missouri. My training was finalized through my score of 4.4 from Oregon State University and I received a certificate as Upper Air Weather Observation Specialist, signed by the Director of the National Weather Service.

For years a plane could ascend from 15 to 20,000 feet into the atmosphere, where the upper records were needed the most, but not in stormy weather. In order to solve this problem the forecasters wanted a meteorgraph and a radio wave transmitter that could be sent up together by attaching them to a balloon. The radio wave transmitter would send down signals giving us weather data from the meteorgraph as the balloon ascends.

The device was actually made and put into operation. It was called a radiosonde. The modulated auto frequency system involved a tiny   transmitter that would broadcast a special frequency radio wave, and attached to the radio transmitter was a small set of measuring instruments, which automatically converts data on the temperature, pressure and humidity into electric impulses. This data was sent by the radio wave transmitter to a receiving and recording set on the ground located inside the Meteorological Observatory at Fort Totten in the Queens.

CONTINUATION OF MY CAREER

AS: UPPER AIR WEATHER OBSERVATION SPECIALIST

My position title as Upper Air Weather Observation Specialist involved critical information given to the pilots, and required working alone on rotating shifts. My shifts began at the Meteorological Observatory at Fort Totten by my preparing the balloon for release. First I filled the balloon with helium. The helium carried the radiosonde instrument up, into the atmosphere.

The balloon weighed about 3 pounds when it was full of helium. I disconnected the balloon from the helium tank and tied a long string using a special knot at the opening of the balloon. The radiosonde instrument was placed at the lower end of the string and the parachute was placed between the balloon and the radiosonde.

The radiosonde or the special frequency radio wave transmitter consisted of a small set of measuring calculations and a baseline check setting, which had to be completed within 20 minutes. The baseline check setting could not be more or less than 3 contacts to reach the state of equilibrium or the radiosonde would be rejected.

After the calculations on the baseline check setting was completed successfully, I attached the radiosonde to the balloon and then it was ready for release. I removed the balloon from the shed and walked quickly outside to the end of the ramp holding the bottom of the string very tightly.

When I arrived at the end of the ramp, I aimed the balloon upward into the atmosphere and released it. As the balloon ascended into the upper atmosphere, the radio wave transmitter automatically transmitted the atmospheric temperature, pressure, and humidity into electric impulses. This data was transmitted to the receiving and recording set located on the ground inside the Meteorological Observatory at Fort Totten. The balloon carried the radiosonde about 1,000 feet a minute into high altitudes frequently ascending over 80,000 feet into the atmosphere.

The lapse rate is determined by the rate of decrease in the temperature of the atmosphere, with the increase of altitude. The rate at which a parcel of air will become cooler or warmer as it gains or loses altitude is approximately 5.5° Fahrenheit per 1000 feet. If the parcel of air is warmer than the environment it will continue to rise and anything greater than the normal lapse rate is super adiabatic. No change in lapse rate with height is Isothermal. Inversions can be considered a lapse rate.

When I released the balloon and looking up into the sky it was like   communicating with God. I quickly returned inside the Meteorological Observatory to the receiving and recording set and began tracking the balloon. As the balloon ascended it would expand and eventually it burst. Then the parachute opens and retards the rate of descent. The slow descent protected the instrument against destruction on impact, and also protected persons or property when it landed. The Military Police on duty at Fort Totten would observe the Meteorological Observatory periodically. No problems ever occurred, therefore my prayers were answered.

When I received the radiosonde recordings of the atmospheric temperature, pressure and humidity from the receiving and recording set I prepared the data for distribution, and transmitted the data by teletype to the National Weather Service Forecast Office, located at the Rockefeller Center office. The forecasters used this information for briefing the pilots before they "took off" from the airports.

Satellites orbiting the earth can measure infrared radiation flowing up into the atmosphere and can also provide information on the temperature and moisture content at various heights in the atmosphere. Satellites that are stationary over one area of the earth can also photograph clouds.

Radar was known and practiced during World War II, by releasing a powerful radio beam. When the beam would strike an object some of the energy reflected back to the receiver and appear as a milky image on the radar screen. The reflection was called an echo. It would give an idea of when the rain would start at a particular area

The Radar Technician sent observations out on the hour. When severe weather was observed on the radar scope the observations were sent again on the half hour. The radar pictures would show the speed and the direction of a storm and the density of the precipitation by the intensity of the light on the radar screen. The radar technician would observe the time that a large thunderstorm might reach the city by the appearance of the echo or milky image on the radar screen.

For many years there was no way to measure the speed of winds above the clouds. A discovery was made by attaching a piece of metal to the helium balloon. Then the radar would follow the receiver after it passed into the clouds and made it possible to record the speed of the winds above the clouds. Some radar of different frequencies can see clouds and airplanes which required the assistance of the electric brain.

CONTINUATION OF MY CAREER

AS: CLIMATOLOGIST

When I completed my assignment at the Meteorological Observatory at Fort Totten, Queens I became the first official Climatologist for the National Weather Service in the New York City area at the Rockefeller Center office, my responsibilities consisted of; vital information given to assist the forecaster and public. I worked alone under my own supervision. My day began as Climatologist by preparing a script. I use the term "script" by repeating a school teachers' description of my work plan for the day. A school teacher was assigned to observe as my assistant for a few weeks. When his assignment was completed a college student replaced him then a high school student. This arrangement of assignments continued until the end of the summer.

My daily responsibilities as Climatologist consisted of maintaining a record of hourly Climatological Data entered on a monthly chart called the Daily Record. I kept a daily record of all Meteorological weather elements used for requested information by the forecaster or public. I also recorded the hourly weather forecasts on tape for broadcast over the radio. I received telephone calls, in office requests and mail regarding weather statistics; and then drove the government car to the New York Meteorological Observatory in Central Park to retrieve the Climatological Data. I replaced the weather charts, recorded the Climatological readings and reset all instruments. I was accompanied by the teacher, college student or the high school student until their assignment ended. They observed and ask questions.

Climatological Data…is data dealing with means either monthly or yearly of various Meteorological elements. The Climatological Data retrieved from Central Park is the official records. It was used for requested information by the forecaster and public. I recorded the daily Climatological information on a chart called the Daily Record. At the end of the month the Daily Record with all records of Meteorological elements were mailed to the National Weather Climatic Center in Ashville, North Carolina, where the information was archived and kept for future requests. The records were printed in the monthly Almanac's and entered in the "Book of Records" which contain records set back in1870 when the first Weather Bureau was under the direction of the Secretary of War.

CONTINUATION OF MY CAREER

AS: CLIMATOLOGIST

I kept a diary to be prepared for requests of the most frequent asked questions in the office, telephone or by mail pertaining to Climatological, Meteorological, Storm and Cloud information.

WEATHER INFORMATION RETRIEVED IN CENTRAL PARK

The Weather Shelter… was located in the compound at the New York Meteorological Observatory in Central Park to protect the dry and wet bulb thermometers from the outside weather and contained:

The dry-bulb thermometer...Gives the temperature of the air.

The wet-bulb thermometer...called wet-bulb because a piece of muslin is placed around the bulb. When the muslin is wet it sends a stream of air from the fan to the wet bulb. As the water on the muslin evaporates it causes the temperature of the wet thermometer to fall. The drier the air, the more evaporation and cooling will take place. The dry bulb thermometer combined with the wet bulb thermometer is called a psychrometer and they determine the content of water vapor. Depression is the difference in temperature between the dry bulb and the wet bulb. Wet bulb and depression equals the dew point. The psychrometer does not give a continuous record.

The Tipping Bucket ...measures the amount of rainfall. It is a small flat bucket, divided in the middle that measures the rainfall. It is balanced on a frame and set under a cone shaped rain spout. When one hundredth of an inch of rain fell into the spout, the rain ran into one side of the bucket, it would tip and repeat until the rain stops. The rain is measured in hundreds of an inch.

The Snow is…a different category of measuring. It is measured by forcing a measuring stick a metal rod or similar depth indicator vertically into the snow until the end rests on the ground surface, not on an ice layer or crushed snow on the ground, and then read the depth indicator on the stick to the nearest inch. The measuring is repeated at several points where the snow is the least by drifting and then I took the reading as the snow depth.

The Sunshine and Wind Instruments were located on the roof of the Belvedere Castle. The wind direction is recorded in the direction that the wind is blowing from. The remote recording anemometer measures the speed of the wind. It has four cups and operates to catch the wind. It is called the Beaufort scale. The stronger the wind the faster the cups would turn. As the cups turned the miles were marked on the cylinder driven by clockwork.

EFFECTS OF THE ATMOSPHERE ON THE HUMAN BODY

ATMOSPHERE…..is the mass of air surrounding the earth.

The Science of Meteorology was born with the invention of the barometer. It is one of the important instruments used by the weather technician. It is used to measure the weight or pressure of the atmosphere. The two principal types are mercurial and aneroid. When the pressure is high, the weather is generally good. The barometer readings were calculated in the National Weather Service Forecast Office.

How Does The Weight of the Atmosphere Affect the Human Body?

When air pressure drops, the balance between the outside pressure and inside pressure in the joints of the human body is disturbed and the result is pain in the joints.

Temperature is…A degree of hotness or coldness, measured on various types of thermometers or some definite temperature scale.

The THI… Temperature-Humidity Index is the combined effects of temperature and atmospheric water vapor on the human body during periods of hot weather.

Human Comfort Spectrum is…Natural air conditioning system is not as effective on days when the relative humidity is high. This is because the high relative humidity lowers the rate of evaporation of perspiration from the skin. Therefore, the rate that heat is taken from our skin is also lowered and our skin is not cooled. When the body feels hot it induces perspiration. As the perspiration evaporates the skin feels cooler, because the process of evaporation removes heat from the skin. Mother Nature furnished us with this type of biological air conditioning.

Heat is…defined as a form of energy and energy is the ability to work. Energy can be in many forms such as the sun is a source of energy. The atmosphere is primary a heat engine.

The wind chill factor…is the opposite of the human comfort spectrum. It numerically describes the chilling effects of the wind during the cold weather. As the wind speed increases, the number which represents the wind chill factor decreases. If the temperature outside is zero and the wind is calm the wind chill factor is zero.

WEATHER MEASURING AND RECORDING INSTRUMENTS:

Ballots Law is when the observer in the northern hemisphere stands with his back to the wind, lower pressure is at his left.

Barograph is a continuous-recording barometer.

Barometric pressure is the same as the atmospheric pressure.

Barometric tendency in aviation weather observations is routinely determined in three hour period.

Barometric tendency is the change of barometric pressure within a specified period of time.

The Beaufort scale is a scale used to measure wind speeds.

Kelvin temperature scale (abbreviated K) is a temperature scale with 0°, equal to the temperature at which all molecular motion ceases.

Maximum wind axis is a line on a constant pressure chart denoting the axis of maximum wind speeds at that constant pressure surface.

Mean sea level is the average height of the surface of the sea.

Mercurial barometer is a barometer in which pressure is determined by balancing air pressure against the weight of a column of mercury and an evacuated glass tube.

Microbarograph is an aneroid barograph designed to record atmospheric pressure changes of small magnitudes.

Maximum thermometer shows the highest temperature reached.

Mean sea level is the average height of the surface of the sea.

Minimum thermometer shows the lowest temperature reached.

Sea level is expressed in millibars. The barometer standard sea level pressure is 1013.2 millibars with 29.92 inches of mercury.

Thermograph is a continuous recording thermometer.

DOCUMENTATION FOR COURT LITIGATION

I was responsible for the documentation of all weather records used for court litigation such as; the Degree Day information and all requests for temperature and weather records requested in the office, by mail or telephone. I requested a specific time for the spring ahead and fall back time change from the Hayden Planetarium to use as the official time for documentation. The time given was at 2 a.m.

Heating Degree Days...

Heating Day commonly called a Degree Day is...a number used to figure fuel requirements for heating. It is a measure of the departure of the average daily temperature from a base value. 65° is used as the base value in order to maintain an indoor comfort level of 70°. This was determined by heating engineers that it is usually necessary to use artificial heat when the main daily outdoor temperature drops below 65°. Fuel consumption varies greatly with the temperature below 65. On a day when the temperature is 20°

below 65, twice as much fuel will be consumed as on a day when the temperature is 10° below 65.

How to Calculate Heating Degree Days...

Heating Degree Days are calculated by subtracting the daily average temperature from 65. For example, if the daily high is 60, and the low 42, the average is 51. This gives a 65 minus 51 or 14 degree days for the day. The daily values can then be accumulated over any time period of interest to the user. The heating degree day season begins on July 1 and ends on June 30.

 Cooling Degree Day is...the less frequency used it is used to estimate the energy requirements, for air conditioning or refrigeration. It is calculated by subtracting the same base value of 65 from the daily average temperature. If the daily high is 92 and the low 70, the average is 81, and we have 31 minus 65 or 16 cooling degree days.

This letter was placed in my personnel folder:

 "Dear Mrs. Durall,

 This is to thank you for providing Climatological Data from the New York Meteorological Observatory at Central Park in such a timely and complete fashion. Attached is a copy of the article, which is being submitted to the local Long Island Newspaper for publication. It is most satisfying to know that conscientious, government workers like you are available to provide this type of information.

Louise Durall, Climatologist for the National Weather Service in Rockefeller Plaza, Manhattan, totaled up the rainfall for 1978: 49.78 inches measured in Central Park. In 1977, she told the Advance Newspaper that the total rainfall measured 40.19 inches".

MY DIARY OF CLIMATOLOGICAL INFORMATION

I kept a diary of vital information pertaining to weather information frequently requested by the public.

CLIMATOLOGY.....is the temperature and moisture in the air. The atmosphere will not hold an indefinite amount of water vapor, it will hold only a limited amount or until no more water vapor can be added, that is called the saturation point or the dew point. A large amount of water vapor can be present in the atmosphere and still have clear weather.

CLIMATOLOGY inquiries:

A Tiny snowflake is made of a single crystal.

Large snowflakes are made of many crystals.

Dew is formed usually at night when the grass and the air close to the   ground cools rapidly below the saturation point.

Hail is from water vapor in the atmosphere, formed by ice crystals gathering the water droplets and growing in size.

Hail and sleet are frozen rain, but snow is not.

Mist is a popular expression for drizzle or heavy fog.

Rain that freezes immediately after falling is called glaze.

Sleet is frozen rain with to which no advantage happens.

Sleet is rain that freezes into clear beads of ice and passes through a layer of cold air before reaching the ground. The beads bounce when they strike the earth.

Snow is water vapor that has been turned directly into ice crystals.

Frost forms when water vapor condenses into droplets and forms ice crystals. The amount of water vapor in the air is called the humidity.

CLIMATOLOGY inquiry:

When a tea kettle boils, steam escapes from the spout and makes a little cloud in the kitchen. The cloud starts towards the ceiling, and suddenly disappears. The steam has turned into an invisible gas called water vapor.

CLIMATOLOGY inquiry:

Water vapor doesn't stay the same. It is always changing. It enters from the bottom of the atmosphere by evaporation from oceans. The winds carry it away. After that condensation occurred. It is carried to cooler layers of the atmosphere and produce rain or snow.

CLIMATOLOGY inquiry:

Water evaporates into the air and becomes a variable constituent of the atmosphere. Water vapor is invisible just like oxygen and other gases. We can measure water vapor and express it as relative humidity and dew point.

CLIMATOLOGY inquiry:

When we open the door of a refrigerator the warm air collects with the cold air on the ice trays. The warm air cools below the saturation point and snow crystals collect on the ice.

CLIMATOLOGY inquiry:

When we put a glass of ice water in a warm room, the glass will collect moisture on the outside. The moisture does not come from the inside of the glass. It comes from the air. When we touch the cold glass the air is cooled below the saturation point and the water vapor comes out of the air in droplets of water on the glass. The same thing happens with nature and it is called dew.

MY DIARY OF METEOROLOGICAL INFORMATION

METEOROLOGY is…study of the free air known as the atmosphere.

Human beings live in an ocean of air. It can't be seen or touched. It has no color or odor. We are totally depended on the invisible ocean of air surrounding us. Our lungs breathe in oxygen to oxidize or burn the waist products in our blood. We take in oxygen from the air and breathe out carbon dioxide. Plants take in carbon dioxide and give oxygen back to the air. A little more than three quarters of the atmosphere is composed of nitrogen. Less than one quarter is composed of oxygen. One part in a hundred is made up of other gases and argon forms only a tiny fraction.

METEOROLOGY inquiry:

If the air were perfectly quiet, the heavier particles and gases would settle close to the earth and the lightest would be found farther out from the earth's surface, but the constant motion of the air near the surface mixes with the gases so that the same proportions exit from the earth's service up to about 45 miles.

METEOROLOGY inquiry:

The lightest gases such as, helium and hydrogen are found at heights above 500 miles. High concentration of ozone and ionized nitrogen, together with smaller quantities of other ionized gases are found in the intermediate levels. Ozone absorption plays a large part in the earth's heat balance, same as the increasing amount of man-made carbon dioxide.

METEOROLOGY inquiry:

Moisture is the most important single component of the atmosphere in producing variations in the weather. Water exists in the atmosphere in three stages; solid, liquid, and gaseous, within the normal range of temperatures. Melting of ice, evaporation of water, and sublimation of ice to vapor are cooling processes. Water takes heat from its surrounding environment during its existing changing of state.

METEOROLOGY inquiry:

Transformation in Meteorology.....is the transformation of water from one form to another such as; solid to ice, liquid or gaseous (water   vapor) to another form are designed by the following five terms

1. Condensation is the change of water vapor to liquid water.

2. Evaporation is the change of liquid to water vapor.

3. Freezing is the change of liquid water to ice.

4. Melting is the change of ice to liquid water.

5. Sublimation is the change of ice to water vapor. Sublimation is also the change of water vapor to ice

METEOROLOGY inquiry:

Clouds are formed from the condensation of water vapor in the atmosphere, vertical motion, moisture and condensation nuclei such as dust. Clouds are the skywriting of the weather because every cloud has a message. They are made of tiny droplets of water or tiny crystals of ice.

METEOROLOGY inquiry:

Dust is a very important element of the atmosphere in making a cloud. It helps to support, the making of a cloud. It is also an important element in the making fog. Fog is a stratus cloud on the ground. Clouds are measured in heights reported above ground level.

METEOROLOGY inquiry:

The drizzle from stable clouds, increase vertical motion droplets and fall out as rain drops. Stable air and stability are atmospheric resistance to vertical motion.

METEOROLOGY inquiry:

The National Weather Service began in 1870, by Congress during the time people were seeking homesteads. They wanted to know where to settle in the country. There were men and women throughout the country keeping weather records on a volunteer basis, in different parts of the country, each building up a local weather picture. The record keeping throughout the country gave the people answers to their questions.

METEOROLOGY inquiry:

Besides being interested in the climate, the National Weather Service became interested in the safety of ships. Storm warnings were needed to advise the nation about storms. The National Weather Service established weather stations, with weather observers to observe the weather regarding storm warnings and ships at sea.

METEOROLOGY inquiry:

Air expands, but not enough to get away from us. Air doesn't expand evenly. The higher up it is the thinner our atmosphere gets and the lower down the denser it gets. That's because air responds to pressure.

METEOROLOGY inquiry:

The air presses on us and also presses down on itself. Air near the earth gets squeezed into a smaller space. It is packed down so that more than half of the atmosphere is in the lower three and one half miles. The first 18 miles above the earth contain 97 percent of the atmosphere. We do not know how far the atmosphere extends. There is no upper surface to our ocean of air. We have learned something about it from shooting stars.

METEOROLOGY inquiry:

Everything that passes near the earth is drawn towards it. The shooting star is a wandering mass of stone or metal that we call meteorites respond to that irritable pull of the earth. They descended through the atmosphere and the friction with the air makes them burn. We can see them burning at heights ranging from 40 to 200 miles. Which means there is air that far up.

METEOROLOGY inquiry:

The aurora is called one of the wonders of the polar region. The aurora borealis is now known to be an electrical action of the sun in the upper atmosphere.

METEOROLOGY inquiry:

The atmosphere is the mass of air surrounding the earth. The troposphere is the first structure of the atmosphere. It is the most important as it lies closest to the earth and is where weather occurs.

The troposphere has a kind of automatic air conditioner. The primary heat pump is the sun. The sun heats the earth's surface and also heats the air in contact with it. Air expands and becomes lighter as it rises. The higher it rises the more it expands, because the pressure around it is steadily lessening. The more it expands, the more it cools. This is an automatic cooling process, which occurs without any loss of heat due to outside causes. The rising air cools automatically about 5 1/2 degrees Fahrenheit per 1000 feet it rises. This is what happens when air rises up the side of a mountain. When it goes down the other side it begins compressing. It warms up at the same rate of 5 1/2 degrees Fahrenheit per 1000 feet and is cooled in rising. This automatic temperature change in rising or falling air is call "adiabatic" warning or cooling. Air does not have to be pushed up a mountain for this adiabatic change to take place. Air rising over heated plains will also cool at the same rate of about 5 1/2°F per 1000 feet rise. As altitude is gained through the troposphere the temperature usually decreases.

Above the troposphere is the stratosphere. Where the troposphere ends and the stratosphere begin is a boundary called the   tropopause. This averages 5 miles above the earth near the poles, and 11 miles above the equator. The stratosphere goes up to about 50 miles.

Above this is the ionosphere extending out to about 650 miles. Here are ionized, or electrified, particles that reflect long radio waves back to the earth. Then above the ionosphere there is the exosphere, which little is known.

METEOROLOGY inquiry:

Air consists mainly of the gases that will not directly sustain life. Air consists of oxygen which all living things need and make up slightly less than 21% of the air. Nitrogen is made up of 78% of the air.

The rest of the gases totaling less than 1% are carbon dioxide, argon, neon, radon, helium, krypton, xenon hydrogen, methane, nitrous oxide, and ozone (a form of oxygen). In addition air contains up to 4% water vapor, also dust and gasses, such as smoke, salt, and other chemicals from sea spray or industry. Air also contains carbon monoxide and micro-organisms.

METEOROLOGY inquiry:

What starts air moving? When the air over the land is warmer than the air is over the water, the warm air expands and becomes light. When the particles are farther apart it starts air moving. Air doesn't always move at the same speed like water flowing from one level to another. The greater the distance and level of the water, the faster the water will flow. The greater the distance of air between the high pressure and the low pressure will cause the wind to blow at a higher speed. The cool air is compact and heavier and presses down heavier than the warm air which makes a difference in pressure and the air will begin to move. The wind blows from sea to land. In the winter it's just the opposite.

METEOROLOGY inquiry:

What is the law of winds: When the air close to the earth tries to move from regions where there is more pressure to regions where there is less pressure, it tries to but through interferences it takes a spiral motion. There is continuous movement in the atmosphere, and continuous circulations of air. Air moves out of and around regions of high pressure and into and around regions of low pressure. It tries to move directly but, is carried off in a curve, or deflected, because the earth is always turning. Wind that starts towards the equator is turned to the right of our hemisphere and to the left of the southern hemisphere. Winds follow directions of troughs causing a lifting.

METEOROLOGY inquiry:

People who live near the sea shore profit by this law in the hot summer weather during the day they get a nice cool sea breeze. The air over the sea is cool and the air over the land is hot. The hot air is pushed upward, and the cool air from the sea takes the place of the hot air. People living as far as 10 to 30 miles inland may feel the breeze. At night it's just the opposite. The land cools more rapidly, then the water. So at night or early in the morning a breeze starts blowing in the other direction; from land to the sea. This is another reason why the sun is behind everything because it heats land and water unequally.

METEOROLOGY inquiry:

Behind all weather, behind the hot summer days, and the cold winter days, behind the winds, behind the rain, and behind the storms is the sun. The sun is not only the source of all light. It is also the source of our weather. The sun is extremely hot. The astronomers tell us that the temperature at the sun's surface must be measured in thousands of degrees. In the interior, the suns heat must be measured in millions of degrees. Such temperatures are too high for the mind to grasp. It is a good thing we are 93 million miles away from the sun and get only a tiny fraction of its rays.

METEOROLOGY inquiry:

As we know our planet spins at a terrific speed around this tremendously hot sun, at the same time, every 24 hours. The earth turns completely around on its axis as it turns each part of the earth's surface is bought towards and then away from the sun. When we say the sun rises and sets, but it is the earth's turning on its axis. This turning brings a change in temperature from day to night. It also causes some of the daily changes in our weather.

METEOROLOGY inquiry:

During winter, summer, spring and fall the earth's path around the sun is not a perfect circle. The sun is not exactly in the center. This means that one part of the path of the earth is nearer the sun then the opposite side. In the winter the earth is nearer to the sun then the summer. It would be warmer in our winter then in the summer if the earth wasn't tilted at an angle. The slanting of the earth's axis makes the difference.

METEOROLOGY inquiry:

In the summer our hemisphere is turned towards the sun. In the winter, when the earth is on the opposite side of the path, our hemisphere is turned away from the sun. In the summer, we are farther away from the sun and its rays reach us more directly. When the sun is shinning directly down on us, it gives us more heat then in the winter because; the sun is low in the sky. In spring and fall, the earth's axis is pointed to the side this gives us less heat then in summer and more then in winter. For this reason, the weather generally is mild in spring and fall. As the earth travels around its own surface and is heated by the sun's rays it will give off heat.

METEOROLOGY inquiry:

The jet stream is a narrow, shallow, meandering river of maximum winds extending around the globe in a wave-like pattern. A second jet stream is not uncommon, and three at one time are not unknown. A jet stream may be as far south as the northern tropics. A jet stream in mid-latitudes generally is stronger than one in or near the tropics. The jet stream typically occurs in a break in the tropopause therefore, a jet stream occurs in an area of intensified temperature gradients characteristic of the break.

The concentrate at winds must be 50knots or greater to be classified as a jet stream.

METEOROLOGY inquiry:

The atmospheric layer just above the tropopause is the stratosphere. The average altitude of this layer is 7 miles at the base and 22 miles at the top. Characteristic of the stratosphere layer is the slight increase in temperature, with height, and the near absence of water vapor and clouds. Occasionally, a strong thunderstorm will break through the tropopause into the stratosphere and, in very rare instances; ice crystals will form a mother of pearl cloud.

METEOROLOGY Inquiry:

With the exception of a substantial increase in the amount of ozone, the composition of the stratosphere is the same as that of the troposphere. Ozone reaches its maximum concentration near the top of the stratosphere. Ozone is important in our everyday lives, because it absorbs most of the deadly ultraviolet rays from the sun.

METEOROLOGY inquiry:

As air masses move out of their source regions, they come in contact with other air masses of different properties. The zone between two different two different air masses is a frontal zone or a front. The fronts are the most exciting places on the map. Most of the bad weather and principal changes in the weather happens along the fronts.

METEOROLOGY inquiry:

Fronts can be either warm or cold. A warm front is a cold air mass retreating and the temperature increasing as the front moves over an area. A cold front is the warm air retreating and cold air advancing and the temperature going down over the area. Air masses are the chief actors in the drama of weather. Highs and lows are secondary. Highs are clockwise air circulation usually with fair weather conditions and Lows counter-clockwise with unsettled weather conditions.

METEOROLOGY inquiry:

An air mass is a very large body of air about 1000 miles or more across, having more or less the same temperature and moisture throughout. If the mass of air stay several days in one place over an ocean, the air mass will be warm and moist. If it should stay several days over cold land, it would be cold and dry. In our country, the air masses move in from the southeast, the south and southwest. Most of them come across southern waters and are moist. They move in ahead of a low. Our cold air masses come down from northern waters and the regions of Canada. They move in north and west of the low center. These warm and cold currents do not mix easily.

METEOROLOGY inquiry:

Lakes and ponds store a great deal of water. Not as obvious as the immense reservoir of water in the polar ice areas, in the glaciers and in snow on mountains and on the cold northern plains during winter. Winter snows in the mountains determine the water supply for irrigation, and for power use. The snow melts, during the spring thaw, and fills the rivers. If spring is late, the melting will be more sudden resulting in floods.

METEOROLOGY inquiry:

Wind is an invisible force, it is air in motion. When the speed increases it becomes stronger and moves faster. The land heats quickly and cools quickly. The water heat slowly and cools slowly. The air close to the land or water is not heated and cooled equally that starts a wind. The suns rays heats the earth. The land gets warmer than the water. The air close to the land becomes warmer than the air close to the water. The warm air expands and becomes light. The cool air is more compact and heavy and there's a difference in pressure. The cool air presses down more heavily than the warm air and that's what starts air moving.

METEOROLOGY inquiry:

Our weather comes to us from the west. And the high pressure tends to move from northwest to the southeast. They often change direction, but in nearly all cases the final result is the same. The highs appear first on the border of the United States in the west or the northwest; and they leave the country from the east or northeast. The high is cool or cold, especially on the east side, where the winds are from the north and northwest.

METEOROLOGY inquiry:

There is a continuous circulation of air, air moves out of the regions of high pressure and into regions of low pressure. It tries to move straight but it is carried off in a curve, or it is deflected. That is because the earth is always turning. Anything loose on the earth's surface which starts to move from one point to another never reach the point because the earth has turned it under and carried it away from the place it was heading. The winds work the same.

METEOROLOGY inquiry:

Any wind that starts toward the equator is turned to the right of our hemisphere and to the left of the Southern Hemisphere. The trade winds are not north and south winds, but northeast and southeast winds because they are turned to the right in our hemisphere and to the left in the southern hemisphere.

METEOROLOGY inquiry:

If you take a jar and pour a little water in it; water will stay at the bottom of the jar. You can't do this with air. You can fill a jar part way with air or with other gases, because gases will take up as much room as you give them, gases expand. Everything in nature obeys the law of nature. The law of gases is that gases expand but like every other form of matter, gases must obey the law of gravity as well as air.

METEOROLOGY inquiry:

The earth pulls the atmosphere towards itself just as it pulls us towards itself. That is the reason why there is still air around the earth. If not the earth would be like the moon, where there is no atmosphere. We can go many days without food, and a few days without water but only minutes without air. We depend on the invisible ocean of air. Our bodies are adjusted to living on its bottom. Our lungs breathe in oxygen to oxidized or burn up the waste products in our blood. We take in oxygen from the air and breathe out carbon dioxide, plants do the opposite. Plants take in the carbon dioxide and release oxygen into the air. The atmosphere obeys the law of gravity as well as the law of expansion. Air expands, but not enough to get away from us. Air does not expand evenly. The higher up the air, the thinner the air will get. The lower the air the more dense the air becomes. That's because air responds to pressure.

Since our bodies are surrounded on all sides by an equal pressure externally and internally. The great pressure causes no comfort. The atmospheric pressure at the earth's surface is the result of the weight of the atmosphere. The greatest pressure is found at the lowest altitudes. As altitude is gained, the atmospheric pressure becomes less. For every 18,000 feet, gain in altitude the pressure is roughly one half of this original value. For the same reason, the pressure is greatest at the bottom of the ocean. And, the pressure is greatest at the bottom of the atmosphere. The average depth of layer is 5 1/2 miles at the poles, 7 1/2 miles over the temperature zones and 11 miles over the equator. Due to the cooling efforts of expansion, the coldest place in the earth's atmosphere is found at the top of the troposphere over the equator. Characteristically, throughout this layer of the atmosphere all the vertical air currents of weather are found.

METEOROLOGY inquiry:

Heat and air movements: the air is heated mainly by contact with the warm earth. When air is warmed, it expands and becomes lighter. A layer of air warmed by contact with the earth rises and is replaced by colder air, which flows in and under it. The cold air is warmed and rises and it is also replaced by colder air.

METEOROLOGY inquiry:

You don't have to heat water over a fire in order to turn the water into a gas. Water continues to turn into gas without the aid of a fire and it dries slowly an example, if you hang wet clothes on a line they will eventually dry which means, the water has risen into the atmosphere and turned into gas. The change is called evaporation. Evaporation takes place from all damp surfaces. Water evaporates from the soil, plants, and from our bodies. The larger the surface, the more evaporation takes place.

METEOROLOGY inquiry:

A breeze in the air helps to carrying moisture away and replaces it with dry air. When the air is cool enough, some of the vapor comes out again in a visible form and condenses. When the air appears in a visible form it is seen as a mist, fog, cloud, dew, frost, rain, snow or hail. Water is the great wonder of nature. It is the only substance that occurs naturally in all forms of matter as solid or liquid.

METEOROLOGY inquiry:

The amount of water vapor in the air doesn't stay the same. It is always changing. It comes from the bottom of the atmosphere by evaporation, mostly from the oceans, because the oceans contain an abundance of water. The winds carry it away sooner or later, as it goes up to the cooler layers of the atmosphere and falls out as rain, or snow. That is the circulation of water it evaporates condenses and comes down as rain or snow fills the rivers and the oceans and starts evaporating again.

METEOROLOGY inquiry:

Rain that freezes immediately after falling is call glaze. The storm that causes it is called an ice storm. Glaze cause great damage to trees and electric wires because, the weight of the ice is so great. Snow has never been water. It is a form of vapor that has turned directly into ice crystals, seldom are two of them exactly alike, they are six sided or six pointed.

METEOROLOGY inquiry:

The rain falls in clouds. It is for the same reason anything falls to earth, and the earth's gravity pulls it. When air is cooling, it becomes saturated and condensation or sublimation begins to form clouds.

METEOROLOGY inquiry:

Saturation results from cooling temperatures, increasing dew point or both. Water droplets in a fog mixed with the smoke may become very dense over a city. This mixture is called smog.

METEOROLOGY inquiry:

Scientists use the word precipitation when referring to rain, snow, hail and other things that come from water vapor in the atmosphere. In the high tops of clouds, there are usually ice crystals which bring precipitation, including extremely small water droplets. The ice crystals gather the water droplets and grow in size. They begin to fall and gather more water droplets whether snow or rain is falling. It depends on the temperatures at the top of the clouds and the air between the clouds, and the earth. If the temperatures are mostly below freezing, snow will fall. If not the ice crystals will melt into raindrops.

METEOROLOGY inquiry:

Modern scientific rain makers' use this basis for their theory. Sometimes they scatter ice pellets or other chemicals into the clouds from airplanes. The rain makers often send chemicals into the sky in the form smoke. They seed the clouds with chemicals. Using this procedure they hope to increase the precipitation and bring more cloud droplets to the earth.

METEOROLOGY inquiry:

Hail stones are born of the thundercloud. When cutting the hailstone in half it is made up of layers that look like the inside of an onion. The layers are composed of ice and snow.

METEOROLOGY inquiry:

After the rain drops are formed in the thundercloud, updrafts inside the thundercloud will lift them up into the freezing zone. If the updrafts of air in the thundercloud weakens and start to fall as they get into the rain level of the cloud, the updrafts will bump into the raindrops and get a coating of rain. Then, the updrafts will lift the raindrops to the freezing heights. The water coating will turn to ice and the hail stones will get another covering of snow before they fall. Sometimes the hail stones are carried up many times. Some hail stones will grow very large and become too heavy for the air to support them and fall to the ground.

METEOROLOGY inquiry:

Sleet is just frozen rain which no adventure happens. The rain freezes into clear beads of ice, because it passes through a layer of cold air before reaching the ground. The beads bounce when they strike the earth.

METEOROLOGY inquiry:

When setting a glass of water in a warm room; the glass sweats on the outside of the glass. This water doesn't come from the inside of the glass. It comes from the air. When touching the cold glass with the hands the air is cooled below the saturation point. The water vapor comes out of the air in droplets and appears as water on the glass.

METEOROLOGY inquiry:

The same thing happens in nature, when it does we call it dew. Dew does not fall. Dew is formed where it is found. At night the grass cools off rapidly. The air close to the ground touches the grass and cools it rapidly also. It cools below the saturation point so the water vapor condenses into droplets on objects such as grass, flowers, bushes and cobwebs. If the temperatures is below freezing the water vapor comes out in the form of ice crystals, this is called frost.

METEOROLOGY inquiry:

Clouds are condensed water vapor and made up of tiny droplets of water or tiny crystals of ice; the particles are so small that a thousand of them set in a row might measure an inch. They are so tiny that the least upward movement of the air will hold them up. They are held up by air currents.

METEOROLOGY inquiry:

Sky cover is referred to as clouds, or other obscuring phenomena, whose bases are approximately at the same level. The layer may be continuous or composed of detached elements. The term layer does not imply that a clear space exists between the layers or that the clouds or obscuring phenomena, composing them are of the same type.

METEOROLOGY inquiry:

We can make a tiny cloud by inhaling deeply through your nose and then exhale slowly through your mouth on a cold day. A little cloud will form around your face from the warmth of your breath. You can imagine you are inside of a cloud.

METEOROLOGY inquiry:

Fog is a cloud close to the ground. The white fleecy clouds that we see in the sky on a summer day are formed differently. They start out as moist air that is being pushed upward.

METEOROLOGY inquiry:

As the air is pushed up it expands, because less air is pressing against it. As it expands it cools. If it cools enough, the saturation point is reached and becomes water vapor. Water evaporates and goes up into the sky and turns into an invisible gas. It is carried away with the winds.

METEOROLOGY inquiry:

Water vapor is in the lower atmosphere, or the troposphere. Sometimes as much as 5% of the total volume of air is water vapor in the form of gas. Generally, there is much less.

METEOROLOGY inquiry:

The amount of water vapor in the air is called humidity. Some of the water vapor condenses or sublimates on condensation nuclei such as dust, haze, smoke or sand and resulting into the formation of a cloud. Water vapor evaporates into the air and becomes an ever-present but variable constituent of the atmosphere. Water vapor is invisible, just as oxygen and other gases are invisible. We can measure water vapor and it is expressed as relative humidity and dew point.

METEOROLOGY inquiry:

Some condensation nuclei have infinity for water and can induce condensation or sublimation even when the air is almost but not completely saturated. As water vapor condenses or sublimates on condensation nuclei, liquid or ice particles will begin to grow. The particles whether liquid or ice does not depend entirely on temperatures. Sometimes the water droplets are too small to fall as rain; therefore the water droplets in the cloud has to join together and increase in size, otherwise rain will not fall.

MY DIARY OF CLOUD INFORMATION

CLOUD FORMATION is…a weather condition of the atmosphere… Cloud forms are endless in their variety. Sometimes we look at clouds and see human faces, animal heads on mountains, birds or fish. There are no two clouds alike and they are constantly changing their shape. The clouds differ from each other same as trees are different from each other. Clouds to almost everyone have some meaning. They differ from each other as trees are different from each other. They give an indication of air motion, stability and moisture. To the pilot clouds are his weather signposts in the sky.

CLOUD DEVELOPMENT inquiry:

A cloud is…a visible aggregate of minute water or ice particles suspended in the air. If the cloud is on the ground, it is fog. When entire layers are cooled to saturation, fog or sleet-like clouds result. Saturation of a localized updraft produces a towering cloud. A cloud may be composed entirely of liquid water, ice crystals, or a mixture of the two.

CLOUD COVERAGE inquiry:

Cloud Coverage is…observed in tenths. One tenth to 5 tenths are known as scattered clouds. 6 to 9 tenths are known as broken clouds.10 tenths coverage are known as an overcast sky or completely covered with clouds.

BASIC CLOUDS inquiry:

There are three basic types of clouds which are: Cirrus, Cumulus and Stratus or high, middle and low clouds. The names of clouds are descriptive of their type and form. They are identified according to the way they are formed.

High Clouds

The Cirrus or high clouds have basis at 20,000 feet or higher and are composed entirely of ice crystals. They look like thin streaks of curls. When the Cirrus clouds thicken another type of high cloud will form called Cirrostratus.

High Clouds

The Cirrostratus clouds form at the same altitude as Cirrocumulus. They are thin sheets that look like fine veils or torn wind blown patches of gauze. They are made of ice crystals and form large halos or luminous circles around the sun and moon. Cirrocumulus clouds generally form at 20,000 to 25,000 feet and are rarely seen.

Middle Clouds

Cumulus or middle clouds meaning accumulation are formed with vertical currents in unstable air. They are the fluffy white heaped up clouds and sometimes called the fair weather cloud.

Middle Clouds

The alto-cumulus auto-stratus and the nimbostratus are three types of middle clouds. Auto-Cumulus clouds are composed of white or gray layers or patches of solid clouds. The cloud elements may have a wave or roll like appearance and some turbulence with small amounts of icing. The Auto-Stratus cloud is thinner and the sun shines through it. The nimbostratus clouds have bases of 6500 to less than 20,000 feet above the ground. The word Nimbus meaning rain cloud is added to the names of clouds which typically produce rain or snow.

Low Clouds

Stratus clouds are low clouds that look like even layers of clouds. They are formed when a layer of warm air over runs a shallow of cold air near the ground. They are like fog and are composed of extremely small water droplets or ice crystals suspended in air.

Thunderstorm Cloud with Anvil

Cumulonimbus clouds are virtually developed clouds of large dimensions with dense boiling tops often crowned with big veils of dense Cirrus, an anvil like form. This cloud is known as the thunderstorm cloud.

It represents violent, vertical movements of the air. They occur as a result of strong uplifting, and always accompanied by lightning and thunder.

MY DIARY OF STORM INFORMATION

In order for a thunderstorm to form the air must have sufficient water vapor, an unstable lapse rate and a strong uplifting, the lower part of the cloud has positive (+) zones surrounded by negative zones (-). When electrical pressure becomes high enough charges between parts of the cloud or between the cloud and the earth are released by lightning. The first lightning strokes are within the clouds.

THUNDERATORM inquiry:

Thunderstorms are…a sign of, "violently" unstable air. Hailstones develop their giant onion like structure by alternately being forced upward by vertical winds inside the thunder cloud to a freezing level and then dropped down to where more water is picked up. Up and down wind motions occur in the clouds but the growth of hailstones results from ice pellets' picking up water in the super cooled middle and upper regions of the cloud. The layers of the giant hailstones result from the differences between the freezing rate and the rate at which water accumulates on the pellets. The path of the hailstones ends when they make their exit from the cloud at the top of the anvil and fall out of the thunder cloud as hail.

THUNDERSTORM/Lightning inquiry:

A lightning discharge is incredibly powerful. Up to 30 million volts at 100,000 amperes, at a very short duration, lightning can not be harnessed or used. The total energy of a major thunderstorm exceeds the energy of an atomic bomb. The sudden tremendous heat from lightning causes, the compression or shock wave called thunder.

THUNDER/LIGHTNING inquiry:

Unstable warm air produces more violent weather. Turbulence is high, and the unstable air sets up ascending air currents creating cumulonimbus clouds and thunderstorms. The precipitation is light and spotty alternating between heavy downpours. As condensation occurs, heat is released, and the air rises violently. At this point, the cumulonimbus stage is reached and might be as high as 40,000 feet.

THUNDER/LIGHTNING inquiry:

Thunder and lightning are…obvious characteristics of the mature stage of the thunderstorm development. The third and the final stage of a thunderstorm development is the dissipation stage. Because of the falling precipitation the thunderstorm downdraft spreads throughout the entire cloud. As the updraft vanishes, condensation ceases and the intensity of all forms of precipitation is reduced. The high-speed winds at very high levels produced the characteristic spreading of the anvil top.

WELL DEVELOPED/THUNDERSTORM inquiry:

A well-developed thunderstorm may exceed upward through the troposphere and penetrate the lower stratosphere. Sometimes the main up draft in a thunderstorm may toss hail out of the top or the upper portions of the storm. An aircraft may encounter hail in clear air at a considerable distance from the thunderstorm, especially under the anvil cloud. Turbulence may be encountered in clear air for a considerable distance, both above and around a growing thunderstorm. Thunderstorms, to the pilot are the most dramatic, most dangerous, and most feared of all weather phenomenon.

TORNADO inquiries:

A tornado may be defined as...a very violent, but short-lived wind. It develops out of the Mammatocumulus cloud in the form of a funnel and heads toward the ground. If it touches the ground it is classified as a tornado.

FUNNEL CLOUD inquiry:

The funnel takes a destructive course on the ground. It sucks up everything in its path. Destructive effects do not come from a forward speed of the storm but from a whirling wind near the vortex which sometimes exceed 300 mph. The funnel cloud descends into a cloud of blowing spray or vapor forms and is picked up by the wind from the surface of a large body of water, beneath the point of the funnel. Finally, the funnel touches the surface and the cloud of spray or vapor takes on the appearance of a column of water. Like the tornado on land, water spouts lasts only for about half an hour.

HURRICANE inquiry:

Hurricanes develop only over ocean areas covered by extremely warm moist air masses. They break up, soon after moving over land. The difference between hurricanes and extra tropical cyclones is the calm "eye" found in the center of the hurricane.

HURRICANE inquiry:

The "eye" of the hurricane is...at the center of the storm. It is a zone of near calm or light breezes, with clear or lightly crowded skies overhead. It averages about 20 miles in diameter. The storm may hit again by full force on the other side of the hurricane.

HURRICANE/EYE inquiry:

It is believed that the eye may be caused by centrifugal force acting on winds at the rim of the eye. The centrifugal force acting on a rotating body doubles when the radius of rotation is cut in half. As air spirals inward toward the center of a hurricane, the centrifugal force increases greatly. The cloudy wall of the eye is where the centrifugal force, exactly balances the pressure forcing air inward to the low pressure center. Friction with the ocean surface will make the whirling air slower and decrease the centrifugal force. The eye is small at the surface, aloft where the wind speed is greater and the centrifugal force is higher. The eye becomes larger and funnel shaped. When hurricanes move over large bodies of land the sauce of moisture which provides energy will discontinue and they began to die. They maintain a more active life span, by traveling over water.

HURRICANE/LIFESPAN inquiry:

Most hurricanes moving northward into temperature zones become involved in extra tropical storms and acquire fronts and distant air masses. Then they lose true hurricane characteristics but still produce damaging winds and heavy rainfall. The average lifespan of a hurricane is nine days. In August, hurricanes last on an average of about 12 days. In July and November hurricanes last about eight days. Calm is the absence of apparent motion of the air.

MY DIARY OF ATMOSPHERIC INFORMATION

BERMUDA HIGH inquiry:

Bermuda High…..in the summer the land area heats more quickly and become areas of low pressure, while the oceans remain relatively cool and become areas of high pressure. These dramatic changes are mostly over the land than over the ocean. This high pressure dominates the northern Atlantic during the summer. The center oscillates between the Azores and Bermuda. When the center oscillates westward to near Bermuda, it is referred to as the Bermuda high. When the high is especially strong near Bermuda, its influence on the Eastern United States is pronounced, with very uncomfortably hot and humid weather.

GREENHOUSE inquiry:

Our global climate is depended on the concentrations of greenhouse gases. If these concentrations increase or decrease our climate will change accordingly. The potential of global warming is tied to carbon dioxide emissions.

GREENHOUSE EFFECT inquiry:

Solar radiation is absorbed and converted into infrared radiation. As this radiates back through the atmosphere to outer space some is absorbed by the greenhouse gases and insulates the earth. This causes the temperature in the troposphere to rise. Various pollutants get in the greenhouse gas content of the atmosphere and more infrared radiation is trapped, leading to a global warming.

COOL GREENHOUSE inquiry:

On a global scale, carbon dioxide, water vapor, and other trace gases in the atmosphere play a role analogous to the glance in a greenhouse. Therefore, they are cool greenhouse gases. The light energy which comes through the atmosphere is absorbed by the earth and converted to heat energy at the earth's surface. The energy now in the form of infrared energy radiates back upward through the atmosphere and into outer space. The greenhouse gases are like a heat blanket, insulating the earth but, allowing the heat to escape eventually. Without this insulation average surface temperatures on earth would be 33° colder and life as we know it now would be impossible.